An ama Management Briefing

The Vital Signs of Effective Packaging Management

David B. Lansdale

amacom
A Division of
American Management Associations

Library of Congress Cataloging in Publication Data

Lansdale, David B
The vital signs of effective packaging management.

(An AMA management briefing)
1. Packaging. I. Title. II. Series: American Management Associations. An AMA management briefing.
TS195.L36 621.7'57 78-25812
ISBN 0-8144-2227-6

This Management Briefing has been distributed to all members enrolled in the Packaging Division of the American Management Associations. A limited supply of extra copies is available at $5.00 a copy for AMA members, $7.50 for non-members.

First Printing

AMA MANAGEMENT BRIEFING

The Vital Signs of Effective Packaging Management

David B. Lansdale

Contents

About the Author

David B. Lansdale was named Director of Packaging for Scott Paper Company in January 1971. He had been manager of staff packaging since December 1957.

Mr. Lansdale joined Scott in 1938 as a purchasing trainee at Chester, Pennsylvania. Three years later he was made staff purchasing agent. In 1952 he was promoted to personnel development manager, and in 1955 he became the Chester plant personnel manager. He was promoted to manager for packaging in 1957. His professional activities include membership on AMA's Packaging Council and the editorial board of *Package Development and Systems* magazine.

A native of Michigan, Mr. Lansdale graduated from the University of Michigan in 1938 with a B.S. degree in mechanical engineering. Mr. Lansdale is an active member of the Packaging Institute and a past president. He is also active in The Society of Packaging and Handling Engineers, where he is a certified packaging professional.

Mr. and Mrs. Lansdale live in Media, Pennsylvania, where they maintain a grove of holly trees that supplies the Christmas market with colorful cuttings in attractive packages.

Introduction

HOW do we know when a packaging operation is being properly managed? This briefing is a guide to help management evaluate the effectiveness of its packaging function. Sales forces have quotas and production units have norms. Packaging, on the other hand, generally lacks a well-defined standard of performance. Primarily a service function, it is frequently spread widely throughout most businesses and affects many people. Its output ranges from developing new ideas to establishing variability controls to helping cut costs in line operations. When packaging works effectively, it is an asset to the business; when it performs poorly, it can sap profits. We will explore the key areas that have to be properly managed in order to make a firm's packaging operation more effective.

Packaging has a wide range of applications. Here, we will concentrate on the packaging of consumer products, because the consumer package is an integral part of the product it is sold with. Consumer packaging is used for identification and protection as well as for dispensing. The cost of a consumer package can be as much as or even greater than the cost of the product it contains. The package is essential to the product's marketability, which makes consideration of the package critical to the manufacturing process. The consumer package's variety of applications also warrants that it receive a high degree of management attention. This study, though based on packaging for consumer products, outlines principles that can be applied to any packaging activity.

Management rarely recognizes the magnitude of total packaging costs, partly because the product receives prime attention. But these costs also escape notice because they are segmented and accounted for in different categories. Material is clearly a packaging cost—but because packaging is handled and stored with other materials, the costs involved are allocated to materials handling and inventory. Sometimes direct labor may be recognized as a packaging cost, but it is more likely to be lumped in with the company's total manufacturing labor account. Similarly, other packaging costs are lumped in with different categories; the costs of operating machinery, support and maintenance labor, and freight and distribution costs.

The true costs of other packaging elements often escape attention. Package design, including the type and amount of material used, is one of these; seldom does excessive or unwarranted packaging show up in the packaging account. Another elusive cost is that of package or product damage resulting from an effort to cut costs by skimping on packaging material or using low quality material. In such cases, paradoxically, normal accounting may show a packaging cost cut when the actual result is a cost increase—one that is rarely attributed to packaging.

Then there is the matter of customer dissatisfaction, which usually results in lost sales. But such dissatisfaction—and the resulting loss of revenue—is rarely attributed to damaged packaging or to the way the package has been made to open, dispense, or protect the product.

Good management of the packaging function is alert to all of these cost factors. This briefing suggests ways to ensure that the hidden costs of packaging are recognized, held to a minimum, and accounted for in their proper categories.

Departments like engineering, purchasing, and accounting are similar from company to company. However, the packaging management function tends to have its own profile in each operation. The service function of packaging demands that it cater to specific needs. Sometimes a single department is made responsible for handling all of a firm's packaging needs, and sometimes their function is splintered and distributed throughout an organization, requiring a coordinating task force to assure unity in its operation. More often, packaging activities are divided by department, according to product group or package division.

Even though practices can vary from company to company, the principles of good packaging management underlie all sound operations. By examining each of the fundamental aspects of packaging and discussing each in turn, this briefing will demonstrate how to apply these principles, whatever the variation may be. By relating the impact of packaging to each activity within the company, we will show how the consequences of inattention to sound packaging principles can be overcome. Excessive costs and how and where they are overlooked will also be discussed.

The ten questions posed in this briefing concern effective management of the packaging function. If you can answer them in the affirmative, it is likely that your firm already has a sound packaging function. Those questions that cannot be answered with an uncompromised affirmative are the areas in which this briefing is designed to help. The questions are:

1. Is there a plan for packaging activities and is it compatible with the firm's goals?
2. Are the packaging needs of those served by the packaging function within the firm known and being carried out?
3. Are the packaging environment hazards known and acknowledged?
4. Is there a system of quality control for packaging?
5. Is the packaging activity properly staffed to do its job?
6. Are the internal and external lines of communication regarding packaging kept open and used aggressively?
7. Is the interface of the packaging function with other departments known and understood?
8. Is there a formal process available and known to all those concerned for the updating of packaging technology?
9. Is there a policy of package exploration and development for the future?
10. Is a broad range of freedom of action available within the reporting relationship?

1

Defining the Packaging Function

THE packaging function may be defined as constituting those activities that assure that optimum packaging is being developed and provided to satisfy marketing, manufacturing, and distribution needs. It is primarily a service activity, responsible for advising and servicing those departments using packages. For example:

- Marketing managers need packages that carry out their sales programs. The package thus becomes a part of the total marketing effort—an effort in which packaging personnel recommend to the marketing manager the best graphics and construction techniques available to do the job.
- Manufacturing requires a consistent supply of packages that perform trouble free at high production rates. Here, the packaging function assists by designing and implementing control standards confluent with this goal.
- Distribution managers must have packages that lend themselves to modern transportation and storage methods. The packaging function must therefore incorporate distribution's requirements into systems and designs that satisfy those needs.

Although a support activity, packaging is still not necessarily subservient to the departments whose needs it serves. It bears the same relationship in its operation to the positions occupied by physicians and their patients or by lawyers and clients. As professionals, packaging experts have to handle their customers within the organization with the same tact and consideration demanded of others providing high-level services.

The packaging function must be large enough to assure that optimum service is rendered throughout the firm. There are both objective and subjective measurements of optimum packaging. The objective criterion is numerical: When less than optimum packaging is used, packaging costs can easily run from 2 to 5 percent higher than in comparable situations that are well managed. Thus, a company spending $10 million annually on packaging might be spending $200,000 to $500,000 more than it has to. On the subjective level, inattention to good packaging practices can result in the loss of consumer and trade goodwill. Unnecessary costs and ill-feeling in the marketplace are matters that no management can afford to overlook. We will explore several of the consequences of such neglect.

The boundaries of accountability for packaging are difficult to determine, because the packaging function interacts with many areas of a business. In addition to marketing, manufacturing, and distribution, it also relates closely with (and at times is a part of) procurement, engineering, and research. Most important is that the packaging operation be large enough to function optimally and equitably throughout the organization. It should not concentrate on any one area to the exclusion of others. Freedom of action in the reporting relationship is essential to optimum packaging. It also is essential that the packaging department have full access to management activities and planning regardless of how the lines for reporting are structured.

THE IMPACT OF PACKAGING

In order to have a better understanding of how the packaging function works at its optimum, the impact of packaging must first be understood. In today's market, it is obviously impossible to get along without packaging; therefore, it would be constructive, at this point, to consider how a product would be gotten to the consumer without a package—which is difficult for most people to imagine. Consumers are aware of packaging but give little thought to it unless it does *not* do what it is supposed to. The package serves them by identifying, protecting, handling, instructing, storing, and dispensing. In this light, it is nearly inconceivable how most products could be marketed without packages. This throw-away element of the product is nevertheless a vital support to its presentation and use, with a broader impact than its sometimes less-than-significant expense might indicate.

On Manufacturing

The production line plays a major part in packaging decisions. Speed and efficiency of manufacture are primary considerations that affect ultimate package design. If a package is inefficient or impractical and fails to fit in with the least-wasteful manufacturing procedures, the entire operation can be adversely affected. Labor, machine time, and materials must be considered when planning packaging. Even though packaging comes at the final phase of manufacturing, it must be considered at every step along the way.

On Distribution

The package is a key link in the modern materials-handling system. A secondary package of the proper size (such as a corrugated box) establishes unit loads that can be locked together to fit on a pallet, on a slip sheet, or into larger containers. Efficient distribution methods also demand that the graphics on packages allow rapid identification by distribution personnel to avoid improper shipments and inventory mix-ups.

Packaging also affects freight costs. Thoughtfully designed packages can allow more units to be placed in a carrier. Efficient carrier handling through careful packaging can result in freight savings that offset the cost of better materials. Careful coordination of package design with distribution needs can minimize the time and effort that carriers spend on handling containers.

Storage space is used to its maximum when packaging is properly designed. Optimal use of storage space can result when packaging material designs are selected with an eye to additional strength and ease of handling. Strong secondary packaging can be stacked higher to take advantage of air space that might otherwise be wasted. The proper sizing of both primary and secondary packages can also lead to a more efficient use of storage space and labor.

Expensive waste incurred by damage and claims not only adds to manufacturing costs but freight and handling as well; such costs are frequently doubled by the necessity for return. Damaged products, when salvaged, are often repacked by labor in the field—labor that is inexperienced and less efficient than plant personnel. Field labor is seldom aware of, and therefore unable to apply, the standards exercised by the manufacturer's highly trained people. This can also result in loss of customer goodwill.

On Marketing

Because the package creates the first impression the consumer receives of the product it contains, it is a key element in the marketing program. The appearance of a package tells a great deal about the product. The care with which a package is presented reflects on the care with which the product was made. Appropriate packaging is an indicator of good management; it places the product in the desired merchandising category.

A properly designed package conveys both the nature and function of a product. It can feature any element: stylishness, strength, value, convenience, color, texture. When inappropriately designed, however, packages can give the wrong impression. A high-priced look, when an economical one would more accurately reflect the product, is as undesirable as a budget-priced look when its appeal should be based on elegance. In short, the package can be both a lure and a deterrent to the customer. It has a big story to tell on its own.

Packages designed to move self-service items act as company salespeople. Their looks convey the image of both the product and the company behind it. Immediate recognition of both is desirable. Package graphics are an extension of the advertising program and provide consumers with reminders of other contacts they have had with the campaign.

The retail outlet is an important factor in well-designed, marketing-oriented packaging. Packages must be designed to fit the shelves on which they are to be displayed or stored. If a package is the wrong size or requires turning and thus will face away from the consumer, potential buyers do not see the panel designed specifically to do the selling. Package width, height, and depth are therefore important considerations, and what appears on each facing has a direct bearing on marketing. If store personnel think a package takes up too much room and they eliminate part of a carefully designed display, marketing goals suffer.

Case count is as vital an element of packaging as any other. If too many packages are put in a case, low turnover may result, causing store managers to stock fewer cases and order a lower gross number of items than they might if each case contained fewer packages. On the other hand, case counts that are *too* low may result in undershelving; empty shelf space is an invitation to competitors. Sales success depends largely on the supply of an item at its outlet, a factor that packaging has a lot to do with.

The materials used for packaging can also strongly influence market positioning. For instance, products put in boxes appeal to different types of consumers than those packed soft. Products typically sold in boxes can lose their potential market when packaged soft because they will be displayed in an unaccustomed place or manner. Conversely, new markets for a product can be built by changing traditional packaging from soft to hard or vice versa. By the same token, the prudent use of different packaging materials can result in wider marketing possibilities—but the wrong choice of material (whether plastic, board, glass, or metal) can act as a sales deterrent.

A vital part of many promotions is good packaging. Not only the graphics tell the story; special displays and multi-pack packaging are also highly significant. They add to the incentive to buy. Promotions that sell products rapidly need easy-to-display packages whose image is clearly defined and linked to the rest of a firm's marketing and advertising program.

On the Consumer

As we have seen, the package in the consumer's hand is an extension of the marketing program. If the package is inviting on the outside but unattractive during or after opening, the consumer may have good reason to wonder if the manufacturer used a similar approach in the product's manufacture. Opening the package is an important act for the consumer. A package that does not open easily can be so frustrating that it prompts the consumer to seek another brand the next time. Improper product fit in the package is another consumer deterrent. An item's being hard to store in or near the area where it is to be used can also affect the decision to buy it again. The method of dispensing a product can sometimes be more important than the product itself. This applies not only to convenience but to the materials used in the package. Ecology-minded consumers want packages that can be reused, recycled, or disposed of with minimum problems. Wasteful or excessive use of packaging materials can therefore influence a growing segment of many markets and be as important as any other packaging features for the consumer.

THE IMPACT OF POOR PACKAGING

We now have a good idea of the impact packaging has throughout most organizations. As obvious as the examples given may be to intelligent managers, their impact is frequently not fully recognized. A more detailed examination of the effects of poor packaging is thus in order at this point. Packaging tends to receive more attention when things go wrong than when it works effectively. Sometimes we know right away when packaging is not working properly, but it can also take weeks, months, or years before we realize that faulty packaging is the reason for poor profits.

At the Point of Sale

When shoppers look at a product display and find that all the items of the brand they want are damaged, they switch brands. Brand loyalties are often outweighed by the desire for neatness and convenience as well as for a specific color, flavor, or design. Damaged packages can result in customers ignoring the intact merchandise they hide. A competitor's product will then carry the market until its supply is exhausted, with sales of the damaged packages lagging far behind (few loyal consumers are willing to overlook damage).

Slow-moving damaged packages cause merchandisers to wait longer before reordering those goods. Removal of damaged products would be a small loss. Rather than suffer this loss, merchandisers wait for the manufacturer to find a solution. (Meanwhile, competing brands are selling.) Outlets may lose a little but the manufacturer loses a lot. The problem of one outlet is multiplied and becomes consequential when package quality in stores throughout a marketing area turns out to be faulty. Sales drop, negating the efforts and expenses involved in advertising and promotion as well.

At the Plant

Careless handling contributes to package damage. However, the problem often lies in packaging design. A package lacking appropriate strength increases the risk of damage. The buying of packaging materials without regard for their quality may be an underlying cause. More careful handling or better machining at some point in the manufacturing process may be one solution to eliminating the damage. The elements of packaging control require continuous attention to make a product look and do its best in the marketplace.

Attention to the manufacturing details of packaging is necessary. Because adequate yardsticks of measurement seldom exist, poor packaging practices at this point are undetected and sap profits. Sometimes sound procedures are forfeited to accommodate cutbacks in the professional attention of packaging.

Skimpy specifications and standards for packaging materials may reduce the administrative expense of the packaging management program. Does this really save money? Purchasing lacks the basis for effective competitive bidding. Perhaps they buy materials at prices that seem lower but something is sacrificed in the way of quality or consistency. The so-called lower cost materials can quickly become the most expensive ones possible because of machine downtime, wasted labor, and spoilage.

At the Planning Stage

Management can unconsciously reverse packaging priorities. For example, take the analysis presented in Figure 1. A company invests $150,000 in a packaging machine that is supposed to fill a product into packages that cost five cents each at the rate of 100 units a minute. The machine is expected to perform 21 hours per day, 300 days each year for five years. Over that period of five years,

Figure I. Potential packaging materials usage – the major cost consideration in machinery selection.

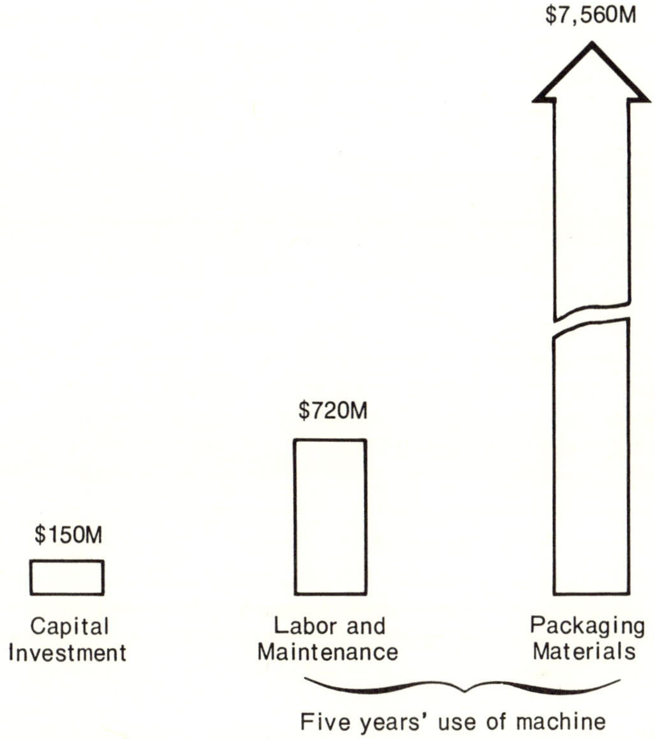

this machine, operating at 80 percent efficiency, will have consumed $7,560,000 worth of packaging materials. This is over 50 times the initial cost of the machine. Labor will have cost $720,000 (assuming there are two operators working at $10 per hour). Clearly, the capital outlay for the machine is the smallest of the three cost figures and packaging materials are the largest cost to be incurred. However, management too often gives more thought to the initial capital expenditure than to long-range operating costs. It might well be that a small increment in the cost of such a machine at the outset could result in a significant reduction in the cost of materials and labor over its life span.

It is a mistake to be inattentive to any of the factors involved in packaging. A soundly based overview of the packaging function includes consideration of both the present and future implications of decision making. An analysis of all the ramifications of packaging should be conducted periodically by management, with proper attention paid to all its possible reverberations. The packaging impact is spread so far and wide in most organizations that it is easy to overlook many of the areas where it affects profits. Its scope and way of making itself felt require that the packaging function be given a different kind of attention from that shown other functions within the organization.

2

The Vital Signs of an Effective Packaging Function

WE will now attempt to answer the questions posed in the introduction. The suggestions in this chapter provide insights into how and why an effective packaging operation functions at its best. The person or groups of people in the organization handling the specialized needs of packaging attend to such functions as structural design, research and development, graphics design, graphics processing, machinery selection, establishment of standards and specifications, procurement, materials management, and quality control. It is not unusual for these activities to report through different management channels within a company. However, the more they are asked to work in unison, the more effective they can be in satisfying the company's packaging needs. These functions all come under the heading of packaging management; it is important that they work in close cooperation with one another. Each of the following subsections concludes with yet another set of questions that are designed to explore further the matters discussed.

A COMPATIBLE PACKAGING PLAN

Is there a plan for packaging activities and is it compatible with the firm's goals? All packaging activities should tie in with both the long- and short-range goals of the company. Whether the company plan is formal or informal, packaging may have an effect on the success of any proposed activity. The impact of packaging is best understood by a packaging expert. The supportive function of packaging is too often overlooked by those who formulate marketing, new-product, manufacturing, or distribution plans. Thus, the most effective way to include packaging in a firm's plans is to involve a packaging specialist in them every step of the way.

Packaging should be allowed to take as much time as necessary in developing its own contribution to planning—and this may amount to as much time again as did the development of the original plan. New-product development, for example, should include a packaging consultant from the start. Manufacturing and distribution must take into account packaging supply and line problems and gear their packaging needs to the future as well as to current usage.

The two ways in which packaging can play a part in planning are: (1) by responding to the ideas and needs specified by management and (2) by generating ideas of its own that become incorporated into the company's overall plan. When packaging is included in all phases of planning, it has an adequate opportunity to contribute to basic inputs. Packaging should be encouraged to shed light on potential problems, contingency actions, and alternative strategies. The packaging specialist should be able to spot areas of sensitivity and give guidance in avoiding costly errors.

There are other time-related reasons for tying package planning in at the early stages of all kinds of planning. It may, for example, take as long to complete shelf-life testing, distribution analyses, graphics development, and product protection testing of the package as it does to develop a new product. Also, the selection of a package in a conference room can be risky. Designs that look good at a marketing presentation may work entirely differently in the marketplace, since a product in quantity looks different from a single, hand-designed prototype.

New-product packaging requires special care because new products receive their greatest scrutiny when they are first introduced. Improper planning can lead to disaster. The early stages of production and distribution are the times, therefore, when new products require that good packaging specifications and standards be met. If the packaging department does not take an active part in the planning process, the new product may not perform properly for its debut.

When the packaging department's function is active as well as reactive, the packaging professional is in a position to generate ideas for company planning. Awareness of changing packaging technology can contribute ideas that lead to improvements in all areas of a business. Packaging should be invited to propose new methods, materials, improvements, and areas of cost savings.

Independent ideas can be expressed only when the packaging department is able to establish a plan of its own. This plan must coincide with the goals of the company and be compatible with the plans of the departments with which packaging interacts. The packaging plan should be reviewed periodically to assure that its goals change as company needs do. (As in most specialized fields, there is sometimes the danger that a stronger loyalty will exist to the ideals of the packaging discipline rather than to fulfillment of the firm's overall goals).

When the packaging function reports through different channels of authority, management has to be assured that each segment ties in with company planning. The various segments of the packaging function should interact to assure that their individual plans mesh with those of the company program.

Questions:

1. Does the packaging function have direct access to the company plan?
2. Does the packaging function contribute to the company plan?
3. Does the packaging function establish plans of its own, and are these plans reviewed regularly to assure their compatibility with the company plan?
4. Do the various parts of the packaging department that report through different management channels cooperate well among themselves?

OVERALL PACKAGING NEEDS

Are the packaging needs of those served by the packaging function within the firm known and being carried out? The primarily service nature of the packaging function makes it necessary that packaging managers be responsible for knowing what areas of the business need to be served, what the needs are, and how those needs can be met. In a marketing-oriented company, the strongest request for packaging services probably comes from the departments in charge of the marketing plan. In other companies, the major request may come from production, distribution, or other departments. Usually, several areas need support simultaneously, especially when those areas are united in a common goal. It is the responsibility of the packaging manager to make sure that the proper balance of attention is given to each area serviced.

If packaging needs are to be understood and met, the packaging function must plan ahead instead of passively waiting for requests to be made. Packaging is an active service as well as a reactive one. Since all departments do not become aware of packaging priorities at the same time, it is incumbent on the resident packaging specialist to anticipate what may be required and make sure that those needs have been incorporated into all levels of planning. Regular contact with each area served should be maintained by a packaging manager to ascertain that the ongoing companywide packaging requirements are properly coordinated.

Planning sessions with the departments served are an essential part of the kind of thinking ahead done by the packaging specialist. When planning sessions are programmed on a regular basis, the packaging function is better able to review changing circumstances that may require its assistance. Also, the department being served is kept up to date on its own packaging requirements. The packaging specialist thus becomes an in-house sales representative, with all the responsibilities that position usually carries. It is a sales position requiring the same care and consideration as any other in handling its important "accounts."

In meeting the demands of activities spread throughout the firm, the packaging people can help top-level management in making sure all projects are compatible with the overall goals of the company. A well-managed packaging function should be able to balance the inputs and demands for attention of its customers while serving the firm's long-range plans.

When two or more departments provide advice and service in a similar area, there may be conflicts of interest. These can be minimized by using the packaging function to help establish interfaces for understanding. (Problem-free periods can be used to negotiate whatever differences may exist, with written guidelines supplied to everyone involved.) This concept will be dealt with at greater length later on.

Just as customer satisfaction is the goal of all good sales representatives, so is it the goal of the packaging specialist. It is not enough to establish action programs during periodic planning sessions; these must be supplemented by regularly held follow-up sessions to assure that the plans are being carried out. Occasionally, it is a good practice to carry the review process up the line of management to gain a broader perspective on the effectiveness of all packaging activities.

Questions:

1. Have the departments, and the people within them, requiring packaging assistance been properly identified?
2. Are the responsibilities of both the packaging department and the areas it serves mutually agreed upon and understood?
3. Does the packaging function plan its activities on a regular basis with the departments it serves?
5. Is the service provided by the packaging function properly balanced throughout the firm?
5. Is there a systematic method of reviewing whether all of the company's packaging needs are being satisfied?

THE PACKAGING ENVIRONMENT

Are packaging environment hazards known and acknowledged?

- It is irritating to everyone when an easy-open spout on a box fails to open readily.
- Discovering a broken handle on an appliance as it is being removed from its package dulls the thrill of acquiring something new.
- A dented can is a sales stopper.
- An overglued package that breaks fingernails results in an unhappy user.

These packaging problems, all of which have a strong impact on product success, arise in the packaging environment. Each of the above situations is the result of faulty packaging. But management is more likely to attribute customer complaints to major line activities (manufacturing, marketing, distribution). The packaging department must be aware of the range of possible conditions that are brought to bear on a package during its entire life. The packaging environment exists at more than just the consumer level. It extends from the package's point of manufacture to its final use and disposal.

Many factors can impinge on the handling and storage of incoming packaging materials. Moisture, temperature changes, humidity, and contaminants in the air can lead to problems later on in the package's life cycle that are not easily traced to its early environment. Poor handling on the manufacturing floor can also generate packaging problems that will not show up till later on. Forming and filling machines can make holes or creases in a package that are hardly noticeable on the production floor but which may result in difficulties at warehouse, store, and customer levels. Potential problems may also be created elsewhere along the line: on conveyors, in individual handling, during storage and shipping, or when items are identified and shelved. An alert packaging department should be aware of all these sensitive spots in the packaging environment.

A package's safety of performance should be audited in each area where it is handled. It is a good idea to institute regular tours of incoming package storage areas. Essential reviews should also be done regularly in the manufacturing area on machine function. An audit system of spot checks should be devised for finished packages as they come off the line that includes opening and scrutinizing them very carefully for damage or malfunctions that are potential causes of problems. Warehouse conveyors and handling practices should also be inspected periodically to make sure they contribute nothing negative to the package's environment.

It is especially important that the environment experienced by the package on its way to the customer be understood. The possible effects of heat, cold, and moisture are factors to be taken into consideration at every step in the package's life. The operators who handle the package in temporary storage areas may not exercise the care necessary to avoid difficulties whose effect is not made apparent at that level. Every time a package is moved, stored, or handled, a situation exists that provides opportunities for problems to creep in.

A regular audit of marketplace conditions is also useful in

controlling damage. Damage in the marketplace is generally reported by the sales force. However, sales' primary responsibility is to sell, not to obtain accurate descriptions of damage. Sales personnel may either underrate problems in the marketplace or exaggerate them disproportionately. A policy of methodical marketplace auditing can provide an effective overview of facts that can direct the packaging specialist's attention to critical environmental areas.

It is not expected that all packages will arrive at the marketplace in perfect condition. Different degrees of damage are acceptable for each product. The packaging specialist should establish an acceptable level of damage for each stage of the packaging environment. In addition to setting standards for the packaging and storage of incoming packaging materials and for their flow through the manufacturing process, the packaging function should also oversee the performance of adhesive applications, regulate handling practices, and specify storage heights. A method of culling damaged packages before they are shipped should be devised as well.

When packaging materials are made or supplied by two or more sources, there should be some way of identifying each supplier in case a problem occurs at any time during the package's life span. The more identifying information in the code, the easier it is for the packaging specialist to isolate the possible sources of problems. Packages made from one position on a sheet, mold, or roll may perform differently from those elsewhere in the same batch. This information should be in the code, in addition to other data, such as time or sequence of manufacture. Coding should include, as well, the filling date and time, manufacturing location, or anything else that may prove helpful in pinpointing problems that may occur long after an item has left the plant.

A good way to achieve efficient control is to keep detailed records of packaging environmental problems. In this way, trends can be observed and corrective action taken before a problem becomes critical. Even when the same problem is manifested in different ways in different parts of the country, as sometimes happens, the packaging specialist's good record-keeping can enable the source of the problem to be tracked. Another benefit of good record-keeping is that it can prevent problems from receiving more attention than they warrant (or less, as the case may be).

Questions:

1. Are the packages and the materials from which they are made coded to facilitate tracing the sources of problems?
2. Have standards of package performance been set for every stage of the package's life cycle?
3. Are all the places a package will travel and the hazards it may encounter known and acknowledged?
4. Is there a method of checking on packaging environment conditions at such key locations as the shipping point or store shelves?
5. Is there an auditing procedure that helps track trends in damage throughout the packaging environment?

MATERIALS QUALITY CONTROL

Is there a system of quality control for packaging? Marketing programs rely on packaging for consistency in the presentation and protection of the product. A steady flow of unvarying basic packaging is required by manufacturing for compatibility with volume production. Distribution counts on strength characteristics whose uniformity ensures safe product handling and storage. The packaging professional must understand these needs so that they can be appropriately communicated to the other areas within the firm.

Color and graphics variations within a group of packages on the shelf indicate to the shopper that some part of a manufacturer's quality-control program is not working as it should. The marketing manager is anxious to avoid this situation, and also wants to be assured that the package's materials and fabrication are consistent with the product it contains. The packaging professional should work with the marketing manager to establish practical ways to make sure the marketing department's standards for ready-to-fill packages are practicable. Objectives and tolerances can be established together for materials, fabrication, and printing. These standards should then be converted by the packaging specialist into terms that can be met and understood by the packaging suppliers and those who have something to do with the package is at each step in the operation. This is how specifications and acceptable quality levels for packaging are set to meet the marketing needs that exist throughout the package's life.

Manufacturing requires packaging materials compatible with the product's manufacturing process. Excessive variations in packaging materials cannot be tolerated; they cause stoppages, slowdowns, and waste. The standards and specifications established by the packaging professional match the packaging materials to the machines that use them. Tolerances for machine variation have to be coordinated with material variances. Overspecification in this area can result in costly waste as easily as underspecification might. Quality requirements that are jointly established by the manufacturing manager and the packaging professional preclude such waste and assure that the best possible production quality standards are met.

Maximum-strength packaging assures lower costs in distribution. It permits fuller use of warehouse space and helps handling systems to gear up for greater speed and efficiency. Strength requirements must be upgraded constantly to meet the demands of increasing speed in carriers. The rigors of distribution require an ample safety factor that can be allowed for when the packaging specialist is on the job. Wide quality variations that affect strength can result in disasters during handling, storage, and transportation—disasters that can be avoided when the packaging professional sets material specifications and performance standards in cooperation with knowledgeable distribution people.

A good quality control program for packaging rests on four steps:

1. Ascertain packaging needs.
2. Establish specifications to meet those needs.
3. Set up process controls with the manufacture of packaging materials to keep them at an acceptable quality level.
4. Audit samples of ready-to-fill packages to check their quality.

Even though the packaging specialist may not take active part in carrying out each step, he or she bears the responsibility of supervising their satisfactory accomplishment within the organization.

Ascertaining Packaging Needs

Different interests within the organization have needs that can vary widely and even be contradictory. Distribution's needs for good unit handling may pose problems for the packaging machines, or marketing may want a shipping container that may prove inconvenient for unit handling. An analysis of needs should consider the basic functions of the package as regards selling, protection, portage, storage, conveying information, and dispensing. When these needs are fully understood in relation to each other and in relation to the requirements of the firm as a whole, the packaging professional is in a better position to establish sound specifications.

Establishing Specifications

The packaging department's broad view of everyone else's packaging requirements makes it the logical group to lead the way in setting comprehensive specifications and standards. With input from other departments, packaging specifications can be broken down in terms of materials, shape, size, strength, color, odor, feel, and function. Standards reflect the objective to be achieved in each of these areas. Once descriptions of packaging quality-control methods and their acceptable tolerance levels have been devised, the packaging function can guide and supervise the full range of packaging requirements as an integral part of the total company picture.

Setting Up Process Controls

Process control is the foundation of quality control. One of the least understood activities of packaging management is the establishment of a system of process controls with the suppliers who manufacture packaging materials. The packaging function has to understand how the supplier's operating methods can be made an extension of its own firm's manufacturing operation. With different suppliers having their own ideas about what constitutes adequate process control, it is the responsibility of the packaging specialist to establish high standards for all suppliers and see that these standards are maintained.

Packaging is not expected to dictate how a supplier should operate his line. It must communicate needs, however, for proper control. Ensuring a consistently high level of manufacture is better than depending solely on intensive inspection procedures to weed out rejects. The supplier's tight control must be matched to the user's needs. When the packaging professional clearly communicates the limits of these needs to the supplier, a degree of surveillance is established to assure that all packaging supplies will remain within allowable limits.

Auditing for Quality

The materials in a given batch must conform to the agreed-upon specifications. Sampling and inspection procedures should set down lines for basic quality control that can be carried out either at the point of origin or at the point of receipt. It is more common for the user to inspect and accept packaging materials on receipt. For long-term suppliers, though, it may be more practical to conduct audits at the manufacturing site, in order to be able to inspect a broader spectrum of samples at their source. That way, unusable materials can be discovered and eliminated before they get to the user.

Maintenance of a cooperative understanding between the packaging supplier and the user is an important function of the packaging professional. Either traveling inspectors or a resident inspector in the user's employ can audit supplies at their source, or the process can be handled by the supplier's own quality-control staff. (There are many advantages to the latter, not the least of which is the understanding and trust it implies in the user-supplier relationship.) Conscientious suppliers also like to perform their own quality audits to keep user rejection levels low. This results in satisfaction on both sides. The packaging professional not only invites the supplier to learn as much as possible about his or her firm's packaging requirements, but also seeks the supplier's help in establishing workable specifications and standards.

Questions:

1. Are the requirements for good packaging known?
2. Do useful and usable specifications exist?
3. Have marketing, manufacturing, and distribution requirements been given appropriate attention?

4. Do the suppliers of packaging materials have adequate process controls?
5. Is there a system of auditing the quality of ready-to-use packages?

STAFFING

Is the packaging activity properly staffed to do its job? As was pointed out earlier, the company-wide packaging costs in a poorly managed operation can run from 2 to 5 percent higher than those in a well-managed operation. A company should feel comfortable in spending about 1 percent of its packaging costs on staffing the packaging function, though it is unfair to relate staff size to the

Table 1. A sample annual packaging activity plan showing personnel requirements.

	Time Required in People-Years		
Function	**Packaging Management**	**Packaging Specialist**	**Technician or Clerk**
Specification development	0.2	0.5	0.5
Needs analysis and trouble-shooting	0.2	0.5	0.5
Package design and testing	0.3	1.0	0.7
Quality surveillance	0.2	0.8	0.8
Administration:			
Planning and control	0.3	0.1	
In-company relationships	0.4	0.3	
Outside relationships	0.2	0.2	
General*	0.3	0.3	1.3
Technology update	0.3	0.3	0.2
Research for the future	0.3	1.0	0.5
Estimated total number of people/years needed**	2.7	5.0	4.5

*Includes record-keeping, letter-writing, and other office work.
**Grand total: 12.2.

overall packaging function cost. Good packaging design and performance at all levels are the objective. Staff size should relate to the amount of its required activity.

An annual packaging plan can help in estimating staffing requirements. The plan may be divided by specialty, with each section reflecting an analysis of the time required. The sum of the segments indicates the total number of people needed each year to execute the functions.

Table 1 suggests the level of expertise necessary to accomplish specific types of activities and the portion of each individual's time that may be required in terms of people/years. It may be noted that the totals in each column in Table 1 are not even numbers of people. The grand total of 12.2 is an estimate, with the personnel in each category being expected to share responsibilities shown under other headings. For example, the specialist would devote some time to management duties, and technicians would also perform some specialist duties. As shown, the department would probably consist of one manager, five specialists, two technicians, two clerical people, and perhaps a specialist in training, for a total of 10 or 11 people.

Having the right people on the packaging staff is as important as having the right number. The expertise needed to carry out packaging specialties requires professionals with a broad knowledge of the field's technology. People who are unfamiliar with the multifaceted aspects of packaging technology may inhibit the firm's ability to branch out in the use of new and improved methods. It is fortunate that the number of graduates trained with specific knowledge of packaging technology is increasing every year to add to the pool of available specialists. Of course, this number is supplemented by other people who have specialized and practical experience and related backgrounds. The qualifications of a packaging staff should include both technical proficiency and a broad base of other knowledge that can assist in problem-solving.

In those instances when people from satellite areas are asked to work on packaging, a program of education should be established to familiarize them with the packaging principles being followed. Courses can be made available in a variety of ways, either inside the company itself or elsewhere (evening schools, university programs, professional seminars or programs, self-study aids). The Packaging Institute offers a home-study course as well as specialized seminars. The Society of Packaging and Handling Engineers has a program that certifies packaging professionals; it includes home study, on-the-job experience, and a comprehensive qualifying test. The American Management Associations offers two- and three-day seminars on packaging topics. Some universities offer concentrated summer programs in packaging.

Planning for future packaging personnel requires special attention by management, whether the packaging function is a small department on its own or is spread throughout the business. Because of the packaging function's frequently dispersed organizational status, it is usually difficult to bring people in at the bottom and have them work their way up. To minimize this problem, the staffing of the packaging function should be planned for in advance.

If management establishes training positions for packaging personnel, a good place to start is in the packaging department itself, in a spot where responsibilities may not be affected by the constant turnovers. The example given in Table 1 shows how a packaging department could operate with 10 or 11 people, with the eleventh position providing a good opportunity for training. Other training spots may be set up in some of the larger departments that use packaging services, such as manufacturing, where a first-hand knowledge of packaging requirements can be gained. In any event, management's staff that is doing the planning for packaging should include estimates for four or five years in advance of current needs.

Questions:

1. Is there an annual packaging work plan that can be used for resource planning?
2. Is the work plan broken down by the number of people/years required to complete tasks?
3. Is the packaging function staffed with qualified professionals with the necessary broad-based knowledge and experience?
4. Does a plan exist for meeting future packaging staff requirements?

LINES OF COMMUNICATION

Are the internal and external lines of communication regarding packaging kept open and used aggressively? Communication lies at the heart of the packaging function, whose major efforts are spent in support of the needs of others. Without strong communication ties, it is difficult for packaging to understand the needs of marketing, manufacturing, and distribution, or to know whether those needs are being satisfied. It is essential that the packaging function maintain good communication—both with internal customers and with groups outside the company. Packaging management is responsible for assuring that these lines of communication are kept open and working relationships remain harmonious.

A system must exist by which the packaging function reports its activities to others within the organization. Periodic meetings of a packaging committee with representatives of other departments can keep everyone abreast of updating in the packaging area. The communication work done by this kind of committee is perhaps as important as any specific work the packaging function performs. Since packaging affects so many activities, it should also be represented in the ranks of management. Among the ways in which the packaging function can keep others informed is via review meetings, progress reports, mini-packaging shows, training seminars, newsletters, and technical reports.

Regular reporting on technical matters is an important avenue of communication through which the packaging department can help other departments avoid duplication. The progressive nature of technical fact-finding requires that a disciplined method of documenting and filing such explorations be maintained within the firm to make this kind of information accessible wherever it may do the most good.

Another vital factor in keeping the packaging function up to date is the necessity to maintain clear lines of communication with people outside the company. Two key links should be mentioned here (the topic will be discussed in more detail later on under *Updating Technology*). The first link is with the firms that provide packaging supplies, because suppliers can be excellent sources of information and help on new technological developments. The second outside link should be with professionals in the packaging community to obtain input on possible new approaches or ideas that have proven workable in other industries.

Questions:

1. Are the lines of communication with the packaging function's "customers" kept open?
2. Are the communications between the packaging function and other service functions kept open?
3. Do the necessary channels exist for making known the activities of the packaging function to those who need them?
4. Are technical activities properly documented and disseminated?
5. Are professional communications outside the company encouraged and actively pursued?

INTERFACES WITH OTHER DEPARTMENTS

Is the interface of the packaging function with other departments known and understood? When several functions provide similar or mutually dependent packaging services, a communication interface among them is needed. Overlapping duties can promote an air of distrust that can cause inefficiencies within the organization. This kind of problem is minimized when interface agreements between departments are developed, understood, and used.

A good interface starts with an understanding, by all parties concerned, of one another's basic purposes. Written statements of purpose are a good beginning. They should include descriptions of why each group exists, whom it serves, what services it provides, and how those services are provided. With everyone understanding these descriptions, it is possible to develop a list of each group's activities and determine respective responsibilities.

The following is a brief hypothetical interface statement between a packaging department and a procurement department:

> Procurement's purpose is to provide advice and service in the acquisition of specified packaging for all departments of this division and to maintain desired inventory levels of that packaging. Packaging's purpose is to develop materials, struc-

tures, and graphics for the packages; it also provides advice and service on supplier selection, development, and utilization of specifications that optimize the cost, quality, service, and performance of packaging objectives for all departments of this division.

From this statement, the following relationships can be understood to exist:

1. Based on inputs from other departments (marketing, manufacturing, distribution), the packaging department will develop, disseminate, and otherwise control the specifications for packaging materials, and provide them to procurement.
2. The procurement department will use the specifications provided by packaging to locate, develop, and select sources of supply; it will meet those standards by having them included in all purchasing contracts.
3. All other contract elements (outside of specifications), such as quantities, price, and logistics, are a procurement responsibility.
4. Packaging will monitor compliance with the specifications included in the purchasing contracts.

In the above example, each department reports to management through a different channel—a situation that affords opportunities for the duplication of effort. Through careful consideration of the purposes and expertise provided by each group in the interface statement, both parties will be able to work together as an effective and harmonious team. The understanding is only as good, however, as the intentions to uphold it. The managers of each department need to review their understanding of the points of agreement on an ongoing basis and resolve minor differences before they become major problems.

Interface agreements between packaging and other departments might be established that would delineate areas of accountability for each as follows:

- *Marketing*—package development and responsibility for accuracy of copy and graphics.
- *Manufacturing*—supply, setup, troubleshooting, machine operations, logistics, rejections.
- *Distribution*—damage control; optimization of package size, shape, and weight; freight classification compliance; storage heights; identification.

- *Quality control*—basic accountability for acceptable quality levels.
- *Engineering*—machinery selection, layouts, scheduling, acceptance trials, control features, and operating manuals.
- *Research*—development of packages for new or improved products.

Questions:

1. Has a statement of purpose describing packaging department activities been developed and accepted?
2. Has a similar statement been developed for the departments that interface with packaging?
3. Have written statements been drafted outlining the relationships that are understood and agreed to by both departments?
4. Are the statements of agreement reviewed on a regular basis?

UPDATING TECHNOLOGY

Is there a formal process available and known to all those concerned for updating packaging technology? Only in recent years has packaging technology been recognized as a field unto itself. People with specialized education in this area are beginning to exert great influence. New materials or new combinations of materials are being introduced at a growing rate, with increased labor costs forcing a high degree of automation on all areas of packaging. Marketing and sales promotion demand wider varieties of innovative ideas, while energy conservation and ecology force compliance with government regulations, presenting challenges that never before existed. In order to meet these growing areas, technology in the field of packaging has expanded at a greater than average rate.

There are many sources that can assist the packaging department in keeping abreast of technological developments. They include:

- *Periodicals—Modern Packaging, Package Engineering, Package Development and Systems, Packaging Digest, Food and Drug Packaging.*

- *Books*—Textbooks such as *Distribution Packaging* by Friedman and Kipnees, *Package Production Management* by Raphael and Olsson, or military handbooks and trade manuals, such as the *Fibre Box Handbook*.
- *Trade shows and seminars*—for sharing knowledge with other experts.
- *Professional organizations*—such as The Packaging Institute, the Society of Packaging and Handling Engineers.
- *Industry field trips*—for observation of companies that produce similar products.
- *Competitive analysis*—Although there is limited exchange between competitors, much can be learned through analysis.
- *Packaging suppliers*—a vital source of information that should always be encouraged.
- *Consultants*—Though not as numerous as in other fields, they can be sources of important information.

Unless packaging professionals disseminate new knowledge properly within the organization, it is worthless. The usual reporting techniques must sometimes be supplemented by other, less orthodox ways of presenting information on new technological developments. Seminars sponsored by the packaging specialist or a supplier and mini-packaging shows can promote a good deal of interest and pass information on in an easy-to-grasp manner. Packaging management should encourage this kind of information flow and help stretch the imagination of personnel in other departments with demonstrations of how new ideas may be effectively applied.

Questions:

1. Does a systematic search for new developments and broader knowledge exist on an ongoing basis?
2. Are more than two of the possible sources of information outlined herein being tapped regularly?
3. Have ways of sharing incoming information been devised to keep interested parties informed?

PLANS FOR THE FUTURE

Is there a policy of package exploration and development plans for the future? A definite portion of the packaging department's time should be set aside for exploration of future packaging needs. With technology changing at a rapid rate, it does not take long to slip behind in the parade of innovations that can enhance the marketing effort or reduce operating costs. The stimulation of such exploration may come either from anticipated future needs within the organization or from developments outside it.

Future company needs should be systematically analyzed by packaging management to make sure all the places are known where packaging can have an impact, currently as well as in the future. For example, in the marketplace, styles in shelf presentations may be changing. Increased labor costs in high-volume stores will demand less package handling and preparation time. Areas of high damage in distribution warrant consideration of alternate methods that can alleviate such problems. High manufacturing costs can also lead to a search for more efficient packaging systems. Opportunities for packaging improvement based on changing needs should be given priority both by management and the packaging department. A combination of ideas from outside the company and from within should be utilized. Methods previously discussed here under *Updating Technology* should be applied creatively.

Packaging suppliers are a good source of knowledge about new technology that can be used, because they usually have a good understanding of their customers' needs. An enterprising packaging department can utilize this expertise and encourage suppliers to take an active role in establishing worthwhile projects for exploration. Some companies ask their suppliers to submit ideas for in-company packaging reviews that are then presented to marketing, manufacturing, and distribution.

Analyzing packaging used by competitors can suggest areas of improvement. A systematic audit of packages found in the market place can reveal the successful use of alternate materials or systems. Ideas for improvement need not be original to be good.

The amount of effort expended by the packaging function on exploration for the future is related to the potential benefits to be derived from exploration. The most popular projects, of course, are finding ways to use materials more efficiently. Cost-saving projects are usually readily understood by management, even to the detriment of other activities that might have greater impact on the organization. But once the initial cream is skimmed, management finds cost-cutting devices progressively harder to find. An aggressive packaging group may be able to generate savings large enough to support its annual budget.

Packaging management needs to encourage those responsible for decision-making in the principal areas it services to seriously consider the possible gains to be obtained from packaging exploration. Resources and time allocations can be determined in proportion to the support given to and by marketing, manufacturing, and distribution. Once projects are established, they should be made part of the company's overall plan.

Questions:

1. Are the future packaging needs of the company reviewed regularly and systematically?
2. Is the new technology relating to these needs being investigated at sources outside the company?
3. Are potential opportunities for exploration documented and assigned priorities?
4. Is the packaging exploration compatible with and integral to the company plan?

FREEDOM OF ACTION

Is a broad range of freedom of action available within the reporting relationship? Packaging should be placed at a reporting level where it can perceive all the crosscurrents that pertain to it. The people who manage the packaging function must have freedom of action in ways that permit them to influence the other areas of the business on which packaging has an impact. Packaging specialists should be able to recommend activities that represent the best overall interests of the company. They are in a position to tune in to many viewpoints along the line. Support of only one point of view is bound to promote opposition from somewhere, which may well develop wasteful competition within the company. Top management should therefore assure the packaging function a broad range of freedom of action in the reporting relationship.

The reporting level of packaging management should be high enough within the organization to allow packaging to participate actively and as an equal in management matters along with those departments it serves. If packaging is expected to contribute to and be part of company planning, it should be involved with planning at a high level. The expectation that packaging serve the needs of other departments can best be fulfilled when it operates on a comparably high level of management with other major functions.

Questions:

1. Is the packaging function so placed that it can give balanced attention to all the company's packaging needs?
2. Is the packaging function located high enough in the organization to enable it to interact effectively with management?

3

Using This Briefing as a Guide

THE packaging activities and responsibilities reviewed in this briefing represent a general cross section of experiences. Undoubtedly there are as many variations in the thoughts presented as there are companies. Obviously, the aim is not to try to be comprehensive, but rather to stimulate overview thinking about the effectiveness of the vital function of packaging within an organization.

Questions have been proposed throughout the text which invite the reader to examine each aspect of the management of packaging. It is suggested that these questions be reviewed honestly and objectively to identify areas that need to be strengthened. On the following pages, for use in the review, is a summary of the questions.

MANAGEMENT GAINS WHEN THE PACKAGING FUNCTION OPERATES WELL

The success of operating the packaging function influences the success of the entire business. Packaging activities, as we have seen, reach into many facets of the business. Packaging directly or indirectly affects costs throughout the company. It usually has a bearing on sales and customer satisfaction. People throughout the business depend on an effective packaging program to maintain their activity. Indeed, packaging may receive very little attention or recognition until it doesn't function as intended. When something goes wrong, attention rises rapidly and the impact is widely felt.

A review of the ten checklist questions reveals where and how

A Checklist for Effective Packaging Management

1. Is there a plan for packaging activities and is it compatible with the firm's goals?

The packaging function should participate in the company plan and contribute to it. Its own plans should tie in with the overall objectives of the company.

2. Are the needs of those who are served by the packaging function within the firm known and carried out?

Those who require packaging services throughout the organization should be identified. The packaging function should participate with them in developing a balanced program of activities.

3. Are the packaging environment hazards known and acknowledged?

The environment of the package should be known throughout its entire lifecycle with problem areas highlighted for attention.

4. Is there a system of quality control for packaging?

Effective quality control requires an understanding of packaging needs in all of its phases, good specifications and standards based on their needs, process controls, and a system of auditing quality levels.

5. Is the packaging activity properly staffed to do its job?

The packaging function should be staffed with properly trained people who work according to an approved work plan. Consideration of future requirements is a part of proper staffing.

widely people are influenced by packaging activities. It starts with the management planning process and carries right through marketing, manufacturing, and distribution operations. It can influence attitudes of people in the business with respect to pride of their product, quality awareness, and even creativity for improvements or better product presentation. Packaging has influence on or relates in some way to almost every department or activity in a business.

6. Are the internal and external lines of communications regarding packaging kept open and used aggressively?

The packaging function is responsible for stimulating and maintaining good communications in all directions with proper documentation of technical information.

7. Is the interface of the packaging function with other departments known and understood?

When departments mutually understand their purposes of existence and document their activities, harmonious teamwork will result.

8. Is there a formal process available and known to all those concerned for updating packaging technology?

A systematic search for new developments and a means of sharing this information with others who need to know is required here.

9. Is there a policy of package exploration and development for the future?

Needs for the future that are developed within the company, coupled with technology development from the outside, form the basis of programming for exploration and development.

10. Is a broad range of freedom of action available within the reporting relationship?

The packaging function should be located in a place where it can give balance and unbiased support to packaging needs throughout the organization. It should be located high enough in the organization to enable it to interact effectively at a management level.

Every place it exists, packaging has some effect on costs. Management usually views as packaging costs those associated with materials, graphics, or capital investments involving packaging equipment. We have seen, however, that packaging does have a bearing on cost-generating factors in each of the main areas of the business. Accounting practices seldom attribute these indirect costs to packaging. For this reason, the total impact of packaging costs is not realized unless management makes a special effort to account for all facets of packaging activities such as those noted in this briefing.

When packaging costs are viewed in total they can be a sizable portion of operating costs. Without a good system of control, 2 to 5 percent of the monies spent on materials can be lost from ineffective practices. A company that spends $10 million annually on packaging could be spending $200,000 to $500,000 more than it should by neglecting good practices in the management of packaging. Cost-saving opportunities are forever present in packaging operations; they need only to be skillfully sought out. The incremental costs of packaging that result from an improperly managed function are difficult to measure. Management awareness of how they can influence total costs is required to make optimum use of packaging.

Because packaging is primarily a service function to line activities in the organization, it is difficult to measure its effectiveness in absolute terms. Unlike the sales or manufacturing departments, it is not possible to set norms or quotas. The output of the packaging function ranges from developing ideas, to establishing variability control, to helping other people accomplish their objectives. When the activity is working effectively it has a positive effect on the business. When it is not working well it can cause problems and sap profits in any area of the business. Management can profit by understanding the key areas of the packaging function and assuring that these are handled in an effective manner.

amacom
A Division of American Management Associations
135 West 50th Street
New York, New York 10020